San Luis Valley Association

## Sunny San Luis

A Complete Description of this Great Agricultural Empire, with Illustrations

of the Methods of Irrigation

San Luis Valley Association

**Sunny San Luis**
*A Complete Description of this Great Agricultural Empire, with Illustrations of the Methods of Irrigation*

ISBN/EAN: 9783337272937

Printed in Europe, USA, Canada, Australia, Japan

Cover: Foto ©berggeist007 / pixelio.de

More available books at **www.hansebooks.com**

# SUNNY SAN LUIS:

A COMPLETE DESCRIPTION

OF THIS

# GREAT AGRICULTURAL EMPIRE,

## WITH ILLUSTRATIONS

OF THE

## Methods of Irrigation.

---

PUBLISHED UNDER THE AUSPICES OF THE

SAN LUIS VALLEY ASSOCIATION

1889.

---

"Through the mossy sods and stones
Stream and streamlet hurry down,
A rushing throng ! A sound of song
Beneath the vault of Heaven is blown !
Sweet notes of love, the speaking tones,
Of this bright day, sent down to say
That Paradise on earth is known."

—*Shelley.*

# BY WAY OF INTRODUCTION.

The writer of this little book desires to put himself on record. He wishes the reader to know that every statement made in the following pages is honestly made, and that this is a plain, un-rhetorical, un-imaginative recital of facts. The space allotted is too limited, and the subject discussed too vast to allow of literary embellishment, although the grandeur of the scenery, the salubrity of the climate, the manifold attractions of the sunny San Luis valley afford a tempting field for glowing description. But those who are looking earnestly about them for a new home, where industry can accomplish a competence, and a competence can grow into enlarged prosperity, do not care to read glowing descriptions; what they want is the truth, the whole truth, and nothing but the truth. They want to know where to find a country whose climate is salubrious, whose soil is productive, whose markets are good, whose schools are well organized, whose society is honorable, industrious, law-abiding and hospitable, and where a poor man can establish a prosperous home or a man with capital find opportunities to increase his store. If such a country exists there are thousands of people who want to know it. They want to know where it is, and they want to know with as much particularity as possible all the facts about it. It is a matter of material interest to them and not an affair of idealism or literary speculation. The writer knows this and he proposes to gratify this natural and proper desire. He knows that the great San Luis valley of Colorado amply fulfills all the requirements enumerated and he has no other wish than to tell in plain language what he knows to be true. He has no especial "axe of his own to grind." His only object is to place the facts before his readers as succinctly as he can and so to fortify his statement with official statistics that even the most skeptical must be convinced that nothing has been exaggerated and nothing set down with unconsidered and incautious enthusiasm. This is plain talk, and it is the intention that this entire book shall be simply a plain talk on a great subject, a live subject, a subject of the most vital interest, a subject that appeals to earnest, honest, industrious men and women, and nothing but self condemnation and public disapproval could be the reward of the writer who should distort or exaggerate the facts. Hoping to have established friendly relations with the reader, and that the following pages will be perused with care and confidence, this book is submitted as an honest description of the Sunny San Luis valley.

SPANISH PEAKS.

# CHAPTER I.

## Early History.

N Rock Creek canon on the western border of San Luis valley are to be seen great cliffs, upon the plain surfaces of which are discernible hieroglyphics and quaint, pictured forms concerning the origin of which there is no history and but little tradition The Spaniards gave the stream the descriptive title of "Piedra Pintada," or Pictured Rock, and it is believed by many that these hieroglyphs were placed there by the Spanish adventurers on the occasion of their incursion into the valley three hundred and fifty years ago. It is known that Coronado came northward from Mexico and swept up the banks of the Rio Grande in 1540, searching for the fabled Madre d'Oro, the mother-land of gold, concerning which the Indians told wonderful stories, readily credited by the gullible and greedy Spainards. It is a curious coincidence that one of the richest gold regions in Colorado has been discovered quite recently in the mountains to the west of the Pictured Rocks, but this discovery has been made some three hundred years too late to sustain either the veracity of the aborigines or to be of any benefit to Coranado and his freebooters.

The Mexican inhabitants of the valley have a tradition among them to the effect that this region was thickly settled before the territory was acquired by the United States, but that, one day, a rumor prevailed among the people that the new government would rob them of their property and enslave their persons and in a single night they gathered their flocks and herds and fled far into the south for safety. This tradition appears to have more of poetry in it than of fact for there is an almost absolute lack of corroborative detail, and the matter rests entirely upon the vaguest kind of evidence. Of course previous to the advent of the Spainards this country was the home of the Indian, and even to this day an occasional degraded descendent of the aboriginal owners of the land may be seen wandering over the plains of his ancestors and begging for tobacco or a draught of "fire water." But leaving the realm of tradition and entering that of history, we find that the first Americans that are known to have entered the valley, were Lieut. Zebulon Pike and twenty-three soldiers, who crossed the plains in 1805, exploring the new country acquired from France by the Louisinia purchase. They were captured on or near the Rio Grande by

6

the Spanish forces, but were finally released. Col. Long came out in 1819, and Cap. Bonneville, of the American Fur Company, in 1832. Gen. John C. Fremont passed through in 1842-44. In 1854 Lafayette Head, since Lieutenant-Governor of Colorado, an American who married a Mexican wife came northward from near Taos, in New Mexico, with a party of followers and founded the present town of Conejos, on the river of the same name, at the lower end of the valley, which is now the county seat of Conejos county. After that settlement, others from New Mexico came still further north and settled along Alamosa Creek, and a few along the Rio Grande river. The Mexicans who still remain in the valley are year

EARLY INHABITANTS.

by year disappearing before the advance of the hardy settlers from the north. They are a quiet, pastoral people, and prefer to associate with those of their own race, rather than the active, bustling American. They however leave behind them the memory of their race, in the names of a few of the towns, counties, streams and mountains, which with their pronunciation and meaning are given below :

San Luis, San Lue—Saint Louis ; San Juan, San Wan—Saint John ; Sangre de Cristo, Sangre de Cristo—Blood of Christ ; Conejos, Kon a-hos —Rabbits ; Rio Grande del Norte—Grand River of the North ; Del

Norte, Del Nor-ta—The North; Alamosa, Alamo-sa—Cottonwood; La Jara, La Ha-ra—Willow; Monte Vista, Monta Vista—Mountain View; Piedra Pintada Creek—Pictured Rock Creek; Hermosilla, Her-mas-ee-ya —Beautiful; Costilla, Kos-tee-ya—Ribs; Huerfano, Wher-fan-oh—Orphan; Carnero, Kar-na-ro—Mutton.

A few families near Saguache, formerly a military post, have resided in the valley over twenty years; but most of those now termed old settlers came only twelve to fifteen years ago. The Denver and Rio Grande Railway was constructed over Veta Pass in 1878, and another branch over Poncha Pass to the coal and iron mines at the northern end of the valley. The line over Veta Pass was extended across the valley to Del Norte and Wagon Wheel Gap in 1881, and another branch was constructed southward to Antonito, and thence across the range to Durango and Silverton; but still the agricultural development of the country was not greatly promoted. The early settlers located upon the natural water courses and devoted their attention chiefly to hay and stock raising, though Mr. Frank Fossett, in his history of Colorado, says, as early as 1879, "there was raised on Saguache Creek, 50,000 bushels of oats, 20,000 bushels of wheat. 50,000 bushels of potatoes, 15,000 tons of hay, besides vegetables in great numbers." The yield of wheat was reported at twenty-five to forty bushels per acre, oats fifty. to eighty bushels, and potatoes two hundred to four hundred bushels.

This in brief is the early history of the San Luis valley. First the home of the nomandic Indian tribes, who hunted over the broad expanses of the meadow-like park, or fished in the sparkling waters of the Rio Grande del Norte and its tributaries, swarming with trout, for which they are still famous; then the field for Coronado's rough riders, whose thirst for gold and blood was insatiable; after this the wonder of the hardy explorers who climbed over its mountain barriers and gazed in surprise and admiration on this "home of millions yet to be;" still later the calm abiding place of the Mexican herdsman, and shepherds, who in their turn are giving way to the increasing advent of sturdy, ambitious and enterprising men, whose brawn and brain are fast transforming this vast valley into a garden of productions.

CHAPTER II.

# Physical Characteristics.

COLORADO is popularly supposed to be a country of mountains, mines and minerals. The general mind does not consider the state in the light of an agricultural country, and yet statistics show that Colorado is in fact one of the best states in the Union for the pursuit of farming industries. To say nothing of the great plains which lie to the east of the Rocky Mountains, Colorado contains four great parks all areable, occupying its longitudinal center, the greatest of which is that of the Sunny San Luis. In these parks are the sources of ten great rivers which radiate and descend without rapids to all the oceans and through all climates. It would be interesting to sketch the characteristics of all these parks, but as that of San Luis is the greatest in extent, is fully prepared to receive a large agricultural population, is supplied with abundance of water and has all the requirements for the making of happy and prosperous homes, we confine our description within its borders. Ex-Governor Gilpin, of Colorado, is an authority on all matters pertaining to this valley, and we cannot better give an idea of its scope and physical characteristics than by quoting what Governor Gilpin has written concerning it: This Park of elliptical form and immense dimensions is enveloped between the Cordillera and Sierra Mimbres. It has its extreme northern point between the two Sierras, where they separate by a sharp angle and diverge, the former to the southeast and the latter to the southwest.

Emerging from the Poncha Pass the waters begin to gather and form the San Luis River. This flows to the south through a valley of great beauty, which rapidly widens to the right and left. On the east flank the Cordillera ascends abruptly and continuously without any foot hills to a sharp, snowy summit. On the west, foot hills and secondary mountains, rising one above the other, entangle the whole space to the Sierra Mimbres.

The Saguache (Sa-Watch) River has its source on the inner (eastern) flank of the Sierra Mimbres, about sixty miles south of its angles of divergence from the Cordillera, and by a course nearly east converges toward the lower San Luis River. It enters upon the Park by a similar valley. These two valleys expand into one another around the mass of foot hills,

9

fusing into the open park, whose center is here occupied by the San Luis Lake, into which the two rivers converge and discharge their waters.

The San Luis Lake, extending south from the point of the foot hills, occupies the center of the Park for sixty miles, forming a bowl without any outlet for its waters. It is encircled by immense saturated savannas of luxuriant grass. Its water surface expands over this savanna during the season of the melted snows upon the Sierras, and shrinks when the season of evaporation returns. From the flanks of the Cordillera on the east, at intervals of six or eight miles asunder and at very equal distances, fourteen streams other than the San Luis descend and converge into San Luis Lake.

The belt of the sloping plain between the mountains and the lake, traversed by so many parallel streams, bordered by grassy meadows and groves of cottonwood trees, has from this feature the name of "Los Alamos." It is sixty miles in length and twenty miles in width. On the opposite (western) side, from the flank of the Sierra Mimbres, similiar streams descend from the west into the lake, known as the Saguache, the Carnero and the La Garita.

The confluent streams thus converging into the San Luis Lake are nineteen in number. The area thus occupied by this isolated lake, and drained into by its converging affluents, forming distinctly one-third of the whole surface of the Park, is classified under the general name of "Rincon."

Advancing onward to the south, along the west edge of the plain, ten miles from the La Garita, the Rio Grande del Norte River issues from the mountain gorge. Its source is in the perpetual snows of the peaks of the San Juan, the local name given to the stupendous culmination of the Sierra Mimbres.

The Del Norte flows from its extreme source due east 150 miles, and having reached the longitudinal middle of the Park, turns abruptly south, and bisecting the Park for perhaps 150 miles, passes beyond its rim in its course to the Gulf of Mexico. All the streams descending from the enveloping Sierras, other than the Alamos, converge into it their tributary waters. On the west come in successively the Pintada, the Rio Gata, the Rio De La Gara, the Conejos, the San Antonio and Piedro. h ese streams, six or eight miles asunder, parallel, equi distant, fed by the snows of the Sierra Mimbres, have abundant waters, very fertile areas of land, and are all of the very highest order of beauty.

Advancing again from the Rincon at the eastern edge of the plain along the base of the Cordillera, the prodigious conical mass of the Sierra Blanca protrudes like a vast hemisphere into the plain, and blocks the

VETA PASS.

vision to the direct south. The road describes the arc of a semicircle round its base for thirty-five miles, and reaches Fort Garland. In the immediate vicinity of Fort Garland the three large streams, the Yuba, the Sangre de Cristo and the Trenchera descend from the Cordilleras, converge, unite a few miles west, and blending into the Trenchera, flow west twenty-four miles into the Rio del Norte. The line of the snowy Cordillera hidden behind the bulk of Sierra Blanca, here again reveals itself, pursuing its regular southeast course and direction. Fourteen miles south is reached the town of San Luis, upon the Culebra River. Seventeen miles farther is the town of Costilla, upon the Costilla River; fifteen miles farther the town of Rito Colorado is reached; eighteen miles farther, on the Arroyo Hondo (between these is the San Christoval) from the Arroya Hondo to Taos is fourteen miles, and twenty miles beyond Taos is the mountain chain whose circle towards the west forms the southern mountain barrier which encloses the San Luis Park in that direction.

The San Luis Park is, then, an immense elliptical bowl, the bed of a primeval sea which has been drained. Its bottom smooth as a water surface and concave, is 9,400 square miles in area. It is watered by thirty mountain streams, which, descending from the encircling crest of snow, converge, nineteen into the San Luis Lake, the rest into the Rio del Norte. An extraordinary symmetry of configuration is its prominent feature. The scenery, everywhere sublime, has the ever changing variety of the kaleidoscope. Entirely around the edge of the plain, and closing the junction of the plain with the mountain's foot, runs a small glacis, exactly resembling the sea beach, which accompanies the junction of the land with the ocean. From this beach rise continuously all around the horizion, the great mountains, elevating their heads above the line of perpetual snow. On the eastern side of the escarpment of the Cordilleras rises rapidly, and is abrupt, on the western side of the crest of the Sierra Mimbres is more remote, having the interval filled with ridges, lessening in altitude as they descend to the plain of the Park.

The colors of the sky and atmosphere are intensely vivid and gorgeous, the dissolving tints of light and shade are forever interchanging— they are as infinite as are the altering angles of the solar rays in his diurnal circuit.

The average elevation of the plain above the sea level is 6,400 feet. The highest peaks have an elevation of over 14,000 feet above the sea. In the serrated rim of the Park, as seen from the plain projected against the canopy, are discernible seventeen peaks, at very equal distances from one another. Each one differs from all the rest in some peculiarity of shape and position. Each one identifies itself by some striking beauty.

From the snows of each one descends some considerable river as well within the Park as outward down the external mountain back.

We recognize, therefore, in the San Luis Park, an immense elliptical basin, enveloping the sources of the Del Norte. It is isolated in the heart of the continent, 1,200 miles from the sea; it is mortised, as it were, in the midst of the vast mountain bulk, where, rising gradually from the oceans, the highest altitude and amplitude of the continent is attained. This Park spreads its plain from 36° to 38° 30', and is bisected by the 106th meridian. Its greatest length is 210 miles, its greatest width is 100 miles, its aggregate appoximate area is 10,000 square miles. *Think of it, six million four hundred thousand acres!*

Such being the geographical position, altitude and the peculiar unique configuration, these features suggest the inquiry into parallel peculiarities of meteorology, mineralogy and the economy of labor.

The American people have heretofore developed their social system exclusively on the borders of the two oceans, and within the maritime valleys of moderate altitude, having navigation and an atmosphere influenced by the sea. To them, then, the contrast is complete in every feature in these high and remote altitudes, beyond all influence of the ocean, and especially continental.

There is an identity between the valley or park of the City of Mexico, and the San Luis Park which ought to be here mentioned. They are similar twin basins of the great plateau classifying together in the physical structure of the continent. Mexico is in latitude 30°, longitude 39°, and 7,500 feet in altitude. The width of the continent here is 575 miles from ocean to ocean, and the divergence of the Cordilleras is here 275 miles, which is here the width of the plateau. At the 39° the continent expands to a width of 3,500 miles from the oceans; the Cordilleras have diverged 1,200 miles asunder, and the plateau has widened to the same dimensions. In harmony with the great expansion of the continent, are all the details of its interior structure. The park of the City of Mexico is but one-sixth contrasted with the San Luis Park. It has an area, including the surface of five lakes, of 1,278,720 acres. Of identical anatomy, the former is a pigmy, the latter a giant. The similitude as component parts of the mountain anatomy is in all respects absolute, as is also true of the other parks which occupy longitudinally the center of the State of Colorado.

The atmospheric condition of the San Luis Park, like its scenery, is one of constant brilliancy, both by day and night—obeying steady laws, yet alternating with a playful methodical fickleness.

There are no prolonged vernal or autumnal seasons. Summer and winter divide the year. Both are characterized with mildness of temperature. After the autumnal equinox the snow begins to accumulate on the mountains. After the vernal equinox they dissolve. The formation of light clouds upon the crests of the Sierras is incessant.

The meridian sun retains its vitalizing heat around the year; at midnight prevails a corresponding tonic coolness. The clouds are wafted away by the steady atmospheric current coming from the west. They rarely interrupt the sunshine; but reflect his rays, imbue the canopy with a shining, silver light at once intense and brilliant. The atmosphere and climate are essentially continental, being uninterruptedly salubrious, brilliant and tonic.

UP THE RIO GRANDE.

The flanks of the great mountains, bathed by the embrace of these irrigating clouds, are clad with great forests of pine, fir, spruce, hemlock, aspen, oak, cedar, pinion and a variety of smaller fruit trees and shrubs, which protect the sources of the springs and rivulets, and among the forests alternate mountain meadows of luxuriant and nutritious grass.

The ascending clouds, rarely condensed, furnish little irrigation at the depressed elevations of the plains, which are destitute of timber but clothed with grass. These delicate grasses, growing rapidly during the annual melting of the snows, cure into hay as the aridity of the atmosphere returns. They form perennial pastures, and supply with food in winter the aboriginal cattle, everywhere indigenous and abundant.

An infinate variety in temper and temperature is suggested as flowing from the juxataposition of extreme altitudes and depressions, permanent snows, running rivers and the concentric courses of the mountains and rivers. Nature is benignant and graceful throughout her whole plan, and is propitious in the working of all her laws and in every element. The longitudinal Sierras receive and absorb the glory of the morning and of the evening sun upon their flanks, the moontide beams upon her summits; they cast no chilling shadow.

Within the bowl of the Park, the rays of the shining sun accumulate. When the sun has set, this heated atmosphere ascends, simultaneously the cooler atmosphere descends from the engirding rim of snow. These atmospheres permeate broadcast, the one the other, through and through, each tempers the other by this play of natural transition. The snows of the altitudes are constantly attacked, and their excessive accumulation defeated; no glaciers form, to enclose the rocks and vegetation as a perpetual tomb. The heat of the concave plain is, in like manner, tempered to a general standard, irrigation and the streams are constantly maintained, vegetation constantly and as uniformly nurtured to maturity. Storms of rain or wind are neither frequent nor lasting. The air is uniformly dry, having a racy freshness and an exhilarating taste. A soothing serenity is the prevailing impression upon those who live perpetually exposed to the seasons. Mud is never anywhere or at any time seen. Moderation and concord appear to result from the presence and contact of elements so various.

The critical conclusion to which a rigid study of nature brings a scrutinizing mind, are the reverse of first impressions. The multitudinous variety of nature adjusts itself with a delicate harmony which brings into healthy action the industrial energy. There is no use for the practice of professional pharmacy. Chronic health and longevity characterize animal life. The envelope of cloud-compelling peaks, the seclusion from the ocean, the rarity of the air inhaled and absence of humidity, disinfect the earth, the water and the atmosphere of exhalations and miasmas.

Health, sound and uninterrupted, stimulates and sustains a high state of mental and physical energy. All of these are burnished, as it were, by the perpetual brilliancy and salubrity of the atmosphere and landscape, whose unfailing beauty and tonic taste stimulate and invite the mental and physical energies to perpetual activity.

## CHAPTER III.

# Salubrious Climate.

**H**EALTH is the first essential for happiness. The Ancients reckoned nothing of more value than "a sound mind in a sound body." The dream of the alchemist was first to find the elixir of life, after that he spent his energies trying to discover how to transmute the base metals into gold. To-day the two great objects of man's desire are health and wealth. When both of these can be found in the same locality the conditions are eminently favorable to a large population. It would be an exaggeration to say that every man who enters the San Luis Valley becomes rich, it would be nothing but the truth, however, to say that here he can find a most excellent opportunity to become rich. The prospects of regaining lost health in this valley are greater than those of regaining lost wealth, for the climate "works" whether the individual does or not, and a man can hardly fail to receive benefit in a physical way, while his financial prosperity must largely depend on his own exertions. The climate of Colorado is especially adapted for the recuperation of exhausted vitality and is a specific for consumption if the invalid avails himself of its benefits in time. The nights are always cool and the summers are never hot or oppressive. Cold stimulates and heat depresses. This is a generally accepted proposition which needs no extended elaboration. The sensations themselves are a good guide, and the colder the air the more stimulating it is. As Dr. Wise expresses it, when introducing the winter climate of the snow-covered regions of the Alps: "A bright sun and blue sky overhead, a clear and quiet atmosphere, distant sounds transmitted to the ear through the still air, combined with the charms of the scenery produce such a bouyancy of spirits that a man is braced and invigorated for almost any exertion." This is an exact description of the climatic characteristics of Colorado, and could not be more accurately stated had the doctor been writing of the Rocky Mountains instead of the Alps. Dr. Charles Denison, of Denver, has made the study of climate a speciality, and in quoting his description of the requirements of a climate favorable to the invalid, especially the consumptive, we are quoting a very thorough exposition of the characteristics of the climate of Colorado. Dr. Denison says: "The clearness or transparency of the air is a decided indication of its purity. It is with the atmosphere as it is with water. The larger the lake, with perfectly clear water, through which one can see to a great depth, the better is the evidence of purity. So a large area, having throughout a similar atmosphere, through which one can see most remarkable distances, and besides, probably be deceived as regards the same, must indicate, as does its coldness, rarefaction and dryness, that the purity is approaching the absolute.

"This increasing purity of atmosphere, that is, the absence of dust or smoke, or of moisture with its attendant infusoria, is a decided feature of elevation, because with each rise of a thousand feet an equivalent stratum of air has been left down below, and, according to Prof. Tindall, each successive stratum contains less and less infusoria.

"A mountainous configuration of country, aside from the benefit of elevation above the sea, has many advantages over a level region. Chief among these are the quick drainage, which allows of no detention of stagnant water; the greater surface of the earth exposed to absorb atmospheric moistures; the many faces of rock, etc., favoring radiation of heat and reflection of light, the element of stimulation both atmospheric and electric, the controlling of severe winds; the variations of scenery, temperature and exposure afforded, and the facility with which one can indulge in the useful "climbing treatment" and pleasurable outdoor activities. When these advantages are compared with the moisture-retaining properties, the sameness, the "siroccos" the trade winds and the "northers" of level regions, there is not much difficulty in choosing between them.

"The changes in the atmosphere, in consonance with the variability of temperature of high climates are in no small degree electrical. There is an increase of electrical tension and an easier and more frequent interchange between the positive electricity of the air and negative quality of the ground and of clouds, so that the condition is decidedly stimulating. This quality in mountainous sections is associated with light showers, especially in summer time, when most needed to clear the atmosphere. The simultaneous whirl of a light or rapid wind often seen in high altitudes, purifies by its substitution of an unused and fresh supply of air for that which is contaminated. Where people crowd together in large numbers, the daily freezing of the air is the only substitute for the movement which is caused by a mild wind. We thus arrive at the conclusion that, in densely settled sections, continuous stillness of the atmosphere is only to be preferred in the freezing weather of winter. In other words the warmer the atmosphere, the more is air movement desirable."

So much for the climate of Colorado in general. Now a few words concerning that of the Sunny San Luis in particular. This valley is situ-

SIERRA BLANCA.

ated at an average altitude of seven thousand five hundred feet, which
would be a source of discomfort were it not entirely surrounded by
high wind-breaking mountains, which with its extremely light, dry air
an l perpetual sunshine, gives it an exceptional combination of conditions pertaining to both northern and southern latitudes, yet wholly unknown to either, namely—a cool all the year climate, the highest summer
temperature in a good house being 84°, with winters so mild that an ordinarily clothed person can read comfortably in the sun on a south fronting
veranda while extremes of heat and cold, (winter nights are cold) are 20° less
appreciable than at sea level.

Ordinarily there is but little snow. With no hail storms (to damage crops),
no tarnadoes, no cyclones, no drouths, no floods, no grasshoppers, no
chinch bugs, and very lew other insects, no possible malaria, with summer nights so cool as to require two blankets for comfort, it is unsur-
passed as a health resort being peculiarly adapted to asthmatics and
consumptives curing in time nearly all who make it their permanent
home, besides being an extra desirable general residence locality. In a
word the climate is far more bracing and exhilerating than that of California and the results much more satisfactory. It must not be supposed
that the climate of Colorado is an equable one, or that there is a distinct
dry and rainy season, as in California and the Pacific Coast. The con-
trary is true. The diurnal range of temperature, as in all high countries,
is great, and there are rains throughout the warm parts of the year and
snows in winter, but both are moderate in quantity.

To render this matter clearer, and to add official weight to the above
statements, the following table is appended:

### YEARLY AND SEASONAL AVERAGES.

Compiled from thirteen years observation.

| Season | Average Tempera-ture. | Average maximum Temperature. | Average minimum Temperature. | Average per cent of Relative Humidity. | Average Rainfall or Melted Snow in inches. | Average Number of Days on which Rain or Snow fell. | Average Number of Clear Days. | Average Number of Fair Days. | Average Number of Cloudy Days. | Average Number of Sunny Days. | Average (in tenths) Cloudiness. |
|---|---|---|---|---|---|---|---|---|---|---|---|
| Spring average | 47.2 | 77.9 | 19.2 | 49.1 | 5.86 | 25 | 33 | 39 | 20 | 81 | 3.0 |
| Summer " | 69.8 | 9.99 | 47.1 | 44.9 | 4.91 | 26 | 37 | 42 | 13 | 89 | 2.4 |
| Autumn " | 29.8 | 80.6 | 19.8 | 45.5 | 2.34 | 15 | 49 | 29 | 13 | 86 | 2.2 |
| Winter " | 49.6 | 62.2 | 7.3 | 54.3 | 1.84 | 15 | 44 | 36 | 10 | 84 | 2.1 |
| Yearly average | 49.1 | 79.2 | 19.7 | 48.4 | 4.95 | 81 | 163 | 146 | 56 | 340 | 2.6 |

JOSEPH J. GILLIGAN,
Observer, Signal Service U. S. A.

A glance at this report, compiled by an officer of the United States Signal service, shows the remarkable fact that 340 out of 365 were sunny days in Colorado.

It is not necessary to add an elaborate argument. The conclusion is self-evident and inevitable. The climate of Colorado, on the whole, presents advantages for the invalid and the pleasure-seeker that cannot fail to command attention.

# CHAPTER IV

# Resources of the Valley.

**T**HE soil of the San Luis Valley is a deep, rich, sandy loam. This loam is of volcanic origin, rich in mineral fertilizers and easily cultivated. With such a soil, plenty of water and almost perpetual sunshine the growth of vegetation is something very little short of phenomenal. With a basis of assured success in agriculture the resources of the valley are exceedingly rich and very various. Many industries are still in their infancy and others have only grown large enough to show what vast possibilities the future holds.

As a horse-raising country the valley is the very best, as there are no fall rains to wash the nutriment out of the grasses, winter feeding is seldom resorted to, while the light air gives a lung capacity, speed and endurance unknown to lower altitudes.

The extraordinarily nutritious grasses and cool climate make the valley one of the most desirable localities for dairying. Blooded stock of all kinds is in much larger supply than in most new countries, and is being constantly imported.

The soil and climate are specially adapted to the grasses, wheat, barley, oats, peas, hops, potatoes and all hardy vegetables. Wheat yields 20 to 50 bushels per acre, barley 40 to 90, oats 30 to 75, peas 30 to 60, potatoes 100 to 400. Potatoes raised here surpass those of almost all other localities in quality, and will soon be shipped to all the cities of the country. Cabbages are as fine as potatoes, and without manure attain a weight of from 10 to 35 pounds each.

Cauliflower is often produced fifteen inches in diameter. Celery makes remarkable crops. Wild hops are larger than cultivated varieties are in other sections, and bid fair to rival those of California and Washington Territory.

Small fruits of all kinds do well; also hardy varieties of apples, cherries, &c., &c.

The surrounding country is full of coal-oil and gas.

Very rich mines are being developed in the mountains to the southwest.

Peas are the corn crop of the San Luis Valley. Sown broadcast they yield from 30 to 60 bushels per acre. Canadian feeders maintain that three bushels of peas make as much pork as four of corn. The pork is certainly sweeter and more healthy. It is less laborious to raise peas than corn. An equal amount of pork can be raised to the acre with that of any Western State,—this without any offense to hog cholera. Hog cholera was never known in the valley.

With tight fences hogs harvest their own peas more than one-half the year. Sown in April early varieties are fit to turn upon July 15th. The autumns are so dry that peas can lie upon the ground without sprouting or being deteriorated in any way as food. The same is true with a majority of winters, nor do rains come to damage them before April. Artichokes grow with great luxuriance. Long, cool springs give three months in which hogs can root artichokes for a living.

) This makes a whole year of human replaced by hog labor. In cheap pork-making Illinois is far behind. And so the great American desert is more than the peer of the richest States.

Hops grow wild on all the streams of the San Luis Valley. They grow larger and are two weeks earlier than any of the cultivated varieties. They have more lupuline and are better adapted to beer-making than any others.

They will yield as well as in the most favored portions of California and Washington Territory. There they have been immensely profitable. They will be so here. Ten acres are enough for a hop yard. This makes a small land outlay sufficient, and gives the many advantages of a dense population. Hop-growing will be one of the many extra desirable industries of the valley.

Gold was discovered in California in 1849, in Colorado in '59, the Comstocks were producing enormously in '69, Leadville was booming in '79, and there is every prospect that the record of the 9s will be fully kept up in the Conejos Camp of the San Luis Valley this year.

One of the last, but not least, resources of the valley is its artesian wells. Artesian water has been found in nearly all localities, while scarcely a boring has proved a failure. Many of the wells have pressure sufficient to force the water from thirty to forty feet above the surface. So far they have been small, four inches being the largest, but present indications will warrant the assumption that they will be available for irrigation. The water is extremely pure, excepting a slight impregnation of sulphate of iron, which renders it a mild tonic. It is very palatable; horses which have drunk it a few weeks will taste no other. Those which have been overworked and generally run down are invigorated and seem to gain new leases of life. Probably it is not assuming too much to say there is not a finer mineral water in the country for constant use in cases of general or special debility. This added to the unparalleled climate, general healthfulness and magnificent scenery of the valley can hardly fail of making it one of the most desirable health resorts in the nation.

We have given in what precedes a condensed statement of the resources of this beautiful and fertile valley. To the readers of this book who are looking for a new home the plain recital of the experiences of a farmer in the valley will doubtless be of especial interest. We, therefore, quote the contents of a paper read before the Farmers' Institute of Monte Vista. This farm spoken of in this report is in no way exceptional, and is a fair average of all the San Luis Valley land, both as to situation, fertility and proximity to market. The author of this paper says:

On January 17th, 1887, I put a pre-emption filing on the N. W quarter of Section 18, Township 39, Range 10, and a timber culture filing on the N. E quarter of the same section, thus securing in one body 320 acres of land, and this is now the subject of this sketch, "My Park Center Farm."

Between January 17th and March 17th, winter as it was, we completed one dwelling and located permanently upon this land.

We went miles beyond any habitation in order to get our land in one body, and seventeen miles beyond where any work had been done on the ditch or canal from which we expected to use water.

The spring of 1887 we fenced eighty acres and put in a small crop of about twelve acres; this was all it was safe to venture on account of water. In the meantime the Prairie Ditch had been so far built as to convey water within three and one-fourth miles of our place, and we built a small ditch or lateral three and one-fourth miles to that point at our own expense.

Owing to imperfections in the construction of the Prairie Ditch its management and our own lateral, we received but little water, and, therefore, lost the principal part of our crops, but were convinced that good results were possible with plenty of water This crop, financially a failure, was rich in experience. We improved our time by clearing land, improving and enlarging our dwelling-house, putting up out-buildings and preparing for a crop the coming season.

Early in the year of 1888 I secured of the San Luis Canal Company water sufficient to irrigate 240 acres of land and retained four shares in the Prairie Ditch. The San Luis Canal Company built its North Lateral in March and April, 1888, and May 1st we were receiving water through it upon our land; therefore, this season, we had plenty of water.

Now for our crop of 1888. Between March 25th and 29th we put in fourteen acres in wheat, seven acres was put in on land plowed one year

23

before and seven acres on land plowed in the fall (six months before), went over the land with a harrow and followed with a drill. Seeded both seven acres in the same way.

DAMING A DISTRIBUTING LATERAL.

We commenced to harvest this wheat August 24th, from the seven acres that lay idle one year (for want of water) we threshed 10,020 pounds, and from the seven acres fall plowed (six months idle) we threshed 7,980 pounds, a total of 18,000 pounds or 300 bushels from fourteen acres.

and a balance in favor of the seven acres first plowed of 2,040 pounds, This wheat was irrigated three times, and as near as I can calculate netted me $12.28 per acre. Sowed 1,200 pounds on fourteen acres and threshed fifteen pounds for every pound sowed. At present prices this wheat would sell at my granary for $1.65 per hundred, or a total of $297 and a net profit of $171 from fourteen acres.

We commenced sowing oats April 16th, finished about the 10th of May, put in about thirty-five acres in all, commenced harvesting oats August 15th, finished October 6th, sold a great many oats from the field at fifty cents per dozen bundles, threshed from twenty-three acres 23,800 pounds, which would bring at present prices at my granary $357, a net profit of $173 from twenty-three acres, with the straw thrown in as a bonus.

April 11th we commenced to plant potatoes; planted at various times until May 15th; planted in all seven acres; commenced digging and selling August 4th; gathering from seven acres a fraction over sixty thousand pounds, which at present prices at my cellar would bring $600. Take from this amount $30 per acre to cover all cost of production, we have $390 net profit from seven acres or $55.72 per acre.

From five acres of barley sowed broad-cast on sod, plowed under and not harrowed, and irrigated in a careless way, sowed late in May, we threshed 5,040 pounds, worth $75.60, a net gain of $35.60 or $7.12 per acre, from this we only expected pasture.

From three acres sowed to peas, put in the same as the barley, we harvested ten tons of forage, worth $5 per ton, $50 in all or $8 per acre net.

We put in one and one-half acres to broom millet, after grubbing the land, sowed the seed broad-cast, plowed and harrowed, threshed 600 pounds of seed, worth five cents per pound, in all $30, or $11 net profit per acre, with a ton of good hay thrown in.

We put in on ten acres, with oats, 150 pounds of alfalfa seed, sowed the seed on the fresh-plowed land, harrowed, then followed with a drill and put in oats. Alfalfa came up well; never saw a better stand, but in my opinion was shaded more than was good for it by the oats, for we noticed where the drill failed to work well and where there were few or no oats the alfalfa made the best growth and is the best now.

Our land is very level, sloping to the east and south (five feet to the mile). We make our small or distributing laterals east and west, about seven rods apart, and our head ditches north and south, eighty rods apart, leaving our land in plots of about three and a half acres. We irrigate grain principally by flooding, but in some cases by seepage or sub-irrigation, but in the latter case find it is best to get the water over the surface about the time the grain is in bloom; this will make it fill and ripen. Sub-irrigation is best for all vegetables. We plowed a piece of land of about four acres the last of June, and the first of July plowed in

OPENING THE WATER GATE.

strips of about two rods, sowed the first plowed to turnips and rutabagas. They weighed from fourteen pounds down. We harvested at least five tons to the acre. We put one acre of this land last plowed, July 3d, to barley, and one acre, July 7th, to oats, the barley headed out well and some of it matured, the oats headed out and made a good growth and some of it matured, making a fine lot of forage and winter pasture. The last four acres was only irrigated by running water in the dead furrows made by plowing the land. We believe it possible to grow large crops of grain by sub-irrigation, but may have to put water on the surface about the time it is in bloom to make it fill well and ripen.

We matured a squash that weighed 51½ pounds, matured water-melons and corn, produced bushels of tomatoes, a large per cent. of them ripening on the vine, grew sweet potatoes of fair size and good quality.

The above is a fair showing of our crop of 1888, all produced on new land that no crop had been grown upon except three or four acres, on which we had water in 1887. This was in our potato and truck patches, but our best potatoes were on new land.

I have told you something of our farm and farming, and wish to state further that my son and myself, either directly or indirectly, have made all the improvements and have done all the work to grow and gather the above-mentioned crops. By this I mean that we have at spare times as-sisted our neighbors and new-comers in building houses, making ditches, head-gates, etc., and that in return they have helped us when wanted with our crops, the balance being in our favor.

We have about 120 acres cleared ready for a crop the coming season, a large per cent. plowed, less than one mile of fence will inclose and leave in three fields our half section of land. Our dwelling and other buildings will show for themselves. What we have done others can do. We have sage brush to contend with (to grub), but no weeds, rocks or adobe. Of course we have to irrigate, but can irrigate more acres of grain in one day than the Eastern farmer can plow of corn, say ten acres per day. We hope one year from now to give a better report of our "Park Center Farm."

A FLUME IN THE MAIN CANAL.

# CHAPTER V

# Farming by Irrigation.

THE eastern farmer thinks that irrigation is either a great nuisance or a great humbug. Then he looks anxiously at the sky and says, " If it don't rain in twenty-four hours my crops will be ruined." The Colorado farmer doesn't need to look at the sky. He looks at his crops and says, "That field needs water ; " then he opens an irrigating ditch and turns on the water. The method of irrigation is simple, the results marvelous. Under this system of irrigation Colorado has become an agricultural state. When all of her irrigable lands are under cultivation she will rank among the great agricultural states. Condensing an able paper read before the State Horticultural Society, we are enabled to present a number of valuable facts, the accuracy of which cannot be questioned :—

The area of land in Colorado at present under cultivation cannot be stated with exactness, owing to the imperfect character of our statistical records, but it is somewhere between one and one and a quarter million acres. It is watered from canals whose aggregate capacity is about forty thousand cubic feet per second, the capacity of these lands being largely in excess of the area irrigated. Estimating the duty of water at one cubic foot per second for eighty acres, the capacity of the canals already constructed is sufficient for the irrigation of more than three million acres. As the area of land susceptible of irrigation is only limited by the water supply, it is apparent that provision has already been made for more than doubling the area at present under cultivation, only awaiting the incoming tide of emigration from the East and South. The exhaustion of the canals already completed does not, however, mark the ultimate extension of our irrigated area. While on some streams in the northern and eastern part of the State the capacity of the canals equals or exceeds the discharge of the stream, there is still a large surplus of unappropriated water and available land in the west and south, the utilization of which is only a matter of time.

Without going into detail of the location and value of these streams a conservative estimate places the amount of land susceptible of being brought under cultivation by irrigation at between four and five millions acres. This is an agricultural area greater than that of either Massachusetts, Rhode Island, Connecticut, New Jersey or Delaware. But the best conception of the wealth and greatness of the State which will result from the development of its irrigation resources, can be had by a study of the irrigation districts of the old world. At present Colorado's irrigated territory surpasses that of France and Spain combined, is about one-fifth that of Egypt, and one-third that of Italy, As the countries last named are the classic lands of irrigation, a few of the results of this phase of agriculture in those districts may not be without interest. The valley of the Nile has an extent of only 6,000,000 acres with an agricultural population of about 2,000,000, yet from the fruits of whose toil are supported the government and people of Egypt. The result of their labors sustain the monasteries with their hosts of religious fanatics, the towns with their bazaars and beggars, the army, the court with its train of officers and dependents, in all a population of 7,000,000 people. And not only is this done, but the interest on their national debt of $515,000,-000 has to be paid out of the products of their teeming soil, the whole furnishing a record of agricultural wealth and resources without a parallel in the most favored regions of abundant rain fall.

The irrigation district in the south of Italy has for investigators of this question, however, the greatest interest, not only because of the beneficent result of irrigation there manifested, but because they are achieved under conditions which in many respects resemble those existing in Colorado. The valley of the Po has an elevation almost equal to Denver, is five degrees to the north of us, its rivers are like ours, fed from the melting ice and snow of its mountain reservoirs, and among its products are found many of our own. The total irrigated area is 3,150,000 acres, having doubled within the past thirty years. The volume of water employed is estimated at 32,000 cubic feet per second, distributed by fifty main canals. The district watered by these canals is the garden of Europe, and in agricultural opulence and commercial stability no country, perhaps, equals it. The provinces Piedmont and Lombardy are the chief contributors to the revenues of the Italian government, the water rentals from the canals alone amounting to $600,000, and the capital of one of the provinces is the commercial metropolis of Italy. A late writer, recounting the results of a tour of observation of the agricultural districts of Europe, says that a town of 20,000 people in the irrigated district of Northern Italy had more evidences of prosperity and thrift than any town of 50,000 he had visited elsewhere. It must be remembered also, in considering this matter, that irrigation in this district is a matter of choice and not of necessity, the rainfall being sufficient to produce crops without artificial watering. The introduction was made under the most discouraging circumstances, and involved an immense outlay. The result has, however, fully justified the expenditure. The introduction of irrigation into these two provinces alone has added one hundred millions to the resources of the Italian government. It has enabled them to withstand the increasing competition of Asia and America, and to add new fields each year to the irrigated lands.

The arid area of the United States embraces nearly one-half of its entire area, or in round numbers about nine hundred million acres. Of this area about two hundred and eighty millions are considered as arable land—that is, land capable of being redeemed and utilized. Of the arable lands nearly thirty-two millions of acres have been redeemed and converted into farms, each of which is capable of supporting a family of five persons. Thus we could have an agricultural population in the arid region of two millions, which would imply a total population of from four to five millions. The present population is something less than two and one-half millions.

The redemtion of the remaining two hundred and fifty millions of acres of arable land is only a question of time, the factors being capital and engineering skill. To a stranger the necessity of irrigation seems an expensive and unprofitable burden to the land and to the cultivator, but experience has proven it to be quite the reverse, and the history of other countries for centuries past also teaches that by it alone is immunity from drouth and famine assured. Crops thus cultivated are not subject to the uncertainty and vicissitudes pertaining to other regions. To specify some of the advantages of irrigation, we have : First, immunity from drouth ; Second, freedom from excessive moisture and flood. Third, the ability to cultivate any kind of plant permitted by the climate, from the aquatic or semi aquatic to that needing the minimum amount of moisture. Fourth, the ability to control in a large measure the growth of the plant, making it early or late as desired, and sometimes growing two crops of the same kind on the same land during the season. Fifth, to control the condition of the soil, making it suitable for the plow and seed. Sixth, to supply certain elements to the soil needed for plant food, namely ; phosphates, sulphate and carbonate of lime, potash and soda salts, nitrogen, magnesia, ammonia, etc., abundantly carried by the water of the ordinary mountain stream. Seventh, to dissolve and wash out certain baneful ingredients of the soil, that frequently exist in such excessive quantities as to destroy plant life. Eighth, the most important advantage in the fertilizing deposit left by the waters, supplying a nutriment to the crop, and adding to and enriching the soil.

The very means of reclaiming the arid land is a constant source of its fertilization. By irrigation the pores of the most sterile land can be

31

filled and compacted by the infiltration of the impalpable silt, and converted into a loam of prodigious fertility. Hence, as a general statement, all lands that can be reached and supplied with water for irrigation are susceptible of cultivation.

Having discussed the merits of irrigation at some length, a few words as to the methods will not be out of place. The name of Orange Judd, the veteran agricultural editor, has become a household word among the

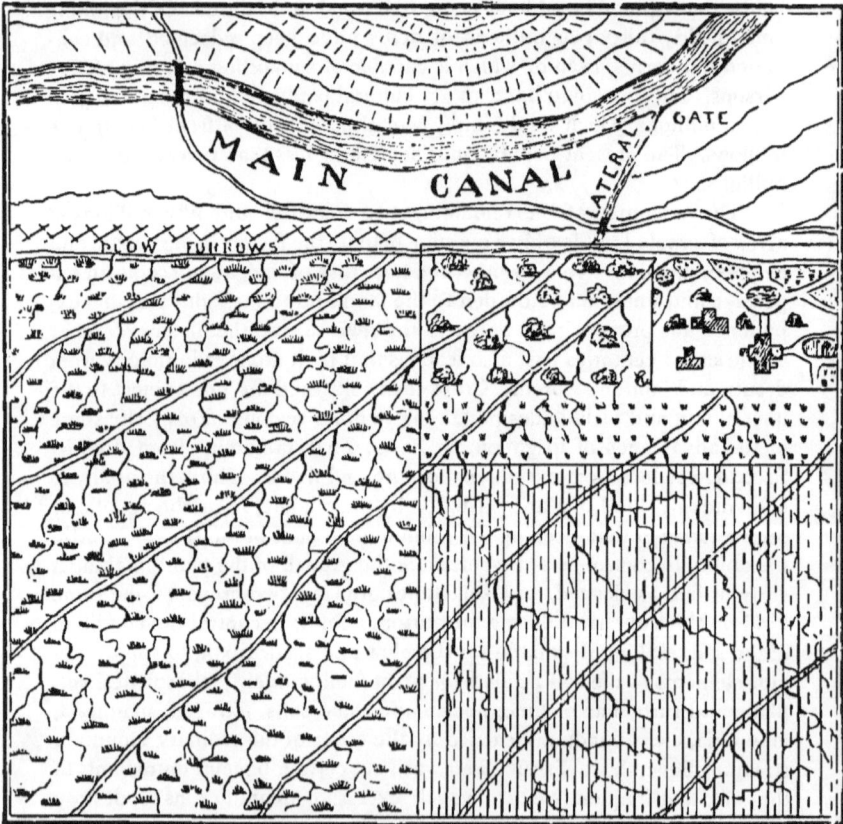

MAP ILLUSTRATING METHOD OF IRRIGATION.

farmers of America. When he describes anything, he does so, accurately and clearly; therefore we cannot do better than to quote what he says on methods of irrigation. Mr. Judd visited the San Luis valley in the summer of 1887, and in his editorial correspondence spoke as follows :—

" Irrigation in the present use of the term means the artificial application of water to the soil, by several methods. One system is shown in the illustration herewith. There is the main canal or 'ditch,' which brings water taken from streams that may be a mile or two, or scores of miles away. A 'lateral' comes out of one side and extends several rods, or even miles to the upper side of a field, into a plow furrow nearly on a level, and the water in this case spreads out each way. From this head furrow very small ones are made with a hoe, or quicker with a small single horse plow. They are run in such direction, required by the lay of the land, as will give them only a slight descent. A hoe or shovelful of earth into the plow-furrow at each entrance of these little ditches keeps them closed. When the land needs water the little gate or sliding board at the canal is raised as far as needed to let in the required amount of water. This is raised or lowered from time to time as seen to be necessary. The large plow furrow being filled with water, the irrigator opens or closes the upper ends of the small furrows by taking out a shovel or hoeful of earth. The operator walks over the field, and where water enough is not flowing out in any place, he with a shovel or hoe clips off a bit of earth from the side of the small ditch or furrow, or stops the flow at any point by throwing in a little soil. In this way he can in an hour or two give an entire field what would be equal to a heavy soaking rain. This may be done so deeply down, one or even two feet, that the growing crop may flourish through the hottest season or drouth without another irrigation.

"Where water goes deep down it is only very slowly evaporated from the surface, while the roots of the crop grow downward so far as to find a good deal of natural moisture in the soil. Usually only two or at most three such irrigations are needed on a wheat crop grown on a soil which is literally a dry ash heap. The number of irrigations and the amount of water at each flowing depend a good deal upon the character of the sub-soil. Some land requires only a single flowing, along in May or June. Sometimes a flooding about the heading-out time will produce very heavy grain kernels. Sometimes the ground is well flooded before the seed is sown and, once or twice afterward unless there is an unusual fall of rain. Most farmers using irrigation rather prefer *no* rain. Having a supply of water in the canal to use whenever needed, they prefer continual hot sunshine which pushes growth forward most rapidly.

In most of the irrigable arid regions these canals are taken out high up a river or stream which is fed by the melting snows of the mountain tops in May, June and July, just the time when plenty of water in the canal is most needed. The canals are carried along with a descent of only 1½ to 2 feet per mile, winding around hills or uneven ground to maintain

33

a uniform grade. If the ground and the stream descend rapidly the canal may thus be carried scores of miles and at its end be 20, 50 100 or more feet above the parent stream. The side canals are taken out at different places and similarly carried over or around uneven land, so that a single main canal may irrigate tens or hundreds of thousands of acres. For example, a canal from a stream in the Rocky Mountains, by following the sides of knolls, valleys and hills, may take water hundreds of miles to supply the parched farms in Eastern Colorado.

MEASURING THE FLOW OF WATER.

"The above is an ordinary method on grain and grass fields which may be flooded all over deeply or thinly. The flooding may be continued during a whole day or more if desired. For corn, potatoes and other crops *in rows*, for fruit trees, etc., one method is to have the rows run *with* the downward incline of the surface, then run one furrow along the upper side of the field to receive the water. A small opening with a hoe against the furrows or hollows between the rows, or opened every second or third or fourth one, allows the water to flow along the furrows and soak into each side of them as long and deeply as desired. The main lateral ditch is usually a permanent one, made by a few plow furrows, not so deep as to prevent easy driving over it. The small field channels are usually obliterated in the general plowing, new ones being made or left in the form and in the places where needed when the crop is put in. A wheat or other grain or grass field is often flooded over its whole surface by openings as needed from the ditch along the higher side. Another method where land is very valuable and permant improvements are desirable, is to run

34

perforated pipe, like drain pipe, 1½ to 3 feet underground, and let water into the heads of them, to soak up into the soil. In this way one has a positive and permanent moisture in the soil "

The San Luis valley is amply provided with irrigating facilities, and the rivers contain more than water enough to irrigate all the land. Over two hundred miles of large canals, with too many miles of laterals for easy computation, are already constructed in the valley at a cost of more than half a million of dollars, while many more are being built. The country is rapidly filling up with water; where four years ago it was fifteen feet below the surface it is now in many places only four or five feet. In closing this chapter we again invoke the high authority of Orange Judd. He makes the statement editorially that farming by irrigation is the most desirable method, and then says:

"This statement will seem singularly inexact, if not foolish, to very many of our readers. It will even be jeered at by the great mass of people between the Atlantic and the Mississippi, simply and only because they do not understand it. We speak from a good deal of examination of the subject last year and this, in Western Texas, in Arizona, in Southern California, in New Mexico, in Colorado, in Utah, etc. 'Seeing is believing,' if the seeing is intelligent and unbiased, and especially if against one's preconceived notions or opinions. The farmer who has a soil containing an abundance of all needed elements, in a proper state of fineness, cannot but deem himself happy if he have always ready at hand the means of readily and cheaply supplying all the water needed by his soil and growing crops just when and in just such quantities as needed. Happier may he still be if besides fearing no drouth, he has no rainfall to interrupt his labors or to injure his growing or harvested crops. And happier still may he be when he knows that he need have few if any 'off years,' and knows that the water he admits to his fields, *at will*, are freighted with rich fertilizing elements, usually far more valuable to the growing crop than any he can purchase and apply at a costly rate—a cost that makes serious inroads upon the profits of a majority of farmers cultivating the worn-out or deteriorated soils in the older States year by year. Fertilizers are already often needed for the most profitable culture of many farms in Iowa, Minnesota, Eastern Kansas and Nebraska, in Missouri, and in all the States east of those named.

" The above are no imaginative statements, but are based upon actual observations—upon a study of the soils, their conditions and wants, East and West, during a long series of years, and upon a firm scientific foundation—so much so that we now unhesitatingly assert that *if allowed to-day to make a choice of a farm for our own future tillage, among the best in the rain fall states or among the best of those only cultivable by irrigation, we should certainly choose the latter.* This is with the under-standing that railways have now made such regions readily accessible to markets, markets as yet better than those in the East, and that there is already so large a class of intelligent, moral people settled in these regions and fast coming into them, as to make society as desirable as in the older States."

# CHAPTER VI

# Towns and Villages.

N agricultural country must have makets not only abroad but at home in order that the exchange of commodities may be easily made and that surplus products may be sold. The law of supply and demand operates in the building of towns as inflexibly as in all other fields of political economy. The demand creates the supply and the supply stimulates the demand. The older towns on the eastern and western borders of the valley were the result of mining industries. The later towns have been created by the new agricultural growth. All the towns and villages now enjoy a permanent and increasing prosperity based on the continued industries of mining, agriculture and stock growing. A condensed sketch of each one of the towns and villages in the San Luis valley will be of special interest to those contemplating a settlement therein.

**ALAMOSA** is situated near the centre of the great San Luis Park on the west bank of the Rio Grande del Norte, and is an important station on the Denver & Rio Grande railroad, 250 miles southwest from Denver. This town was the centre of great commercial activity at the time when the Denver & Rio Grande railroad was built through the valley. At Alamoso it halted in construction for a long time. The town then became the forwarding point for all southern Colorado and New Mexico. But presently the railway moved southward and westward, and Alamosa settled down into a quiet yet prosperous place, with a local agricultural population to back it, and the headquarters of the second division of the railway which extends to Espanola with direct rail connection beyond to Santa Fe. From Alamosa a branch line runs to Wagon Wheel Gap, passing through Monte Vista and Del Norte *en-route*. The fact that Alamosa is a junctional point and an eating station brings to it a great deal of railroad business and makes it the stopping place for a large number of tourists. At present it has a population of 2,000, excellent Schools, three churches, and two newspapers. The streets are wide, handsome brick buildings supply the needs of bank and all the leading business houses. Many of these structures would do credit to a city of much larger size and population. Alamosa is the site of division headquarters of the Denver & Rio Grande railroad and machine shops, round-house and other commodious buildings of brick have been erected. Natural gas has been discovered near the town and the discovery promises to develop into a large source of prosperity for Alamosa. The agricultural resources of the outlying country have been amply discussed in this book. There is room in this great valley of San Luis for thousands of prosperous and happy homes. It is being rapidly settled by a good class of people and will no doubt in a few years become one of the most important agricultural districts in the United States. With these wonderful agricultural and pastoral resources Alamosa is assured of a bright and prosperous future. The town is now well supplied with stores of all kinds, some of which carry large stocks of goods.

**ANTONITO** is a thriving town in the San Luis Valley, situated 279 miles from Denver at the junction of the Silverton and the Espanola Branches of the Denver and the Rio Grande railroad. Stock raising and agriculture occupy the attention of the surrounding population. Antonito has a fine stone depot and possesses a number of substantial business blocks. It is the station for Conejos, one mile distant, for Manasa, a large and prosperous Mormon settlement, in which polygamy is not practiced, 8 miles distant, and for Rafael 4 miles distant. Its position in the heart of the San Luis Valley insures it a generous and constantly increasing support from agricultural and pastoral industries. Being the junctional point of the Denver and Rio Grande railroad's New Mexico and San Juan branches gives it a large railroad business. Fine fishing can be found near Antonito. Antonito itself is a modern town with all the life and push of the American, full of business and enterprise.

**CONEJOS,** an old Mexican town of curious interest, is the county-seat of Conejos county, on the Conejos river, in the San Luis Valley, 279 miles southwest of Denver and 29 mile south of Alamosa. Railroad connection is one mile south of the town at Antonito on the Denver & Rio Grande railroad. Tourists will do well to stop at Antonito and visit Conejos, which is the most accessible town of the typical Mexican character in Colorado. Here may be found the plazas, churches and ancient adobe houses peculiar to the early civilization of the Spanish. Aside from its historical interest the town has a good business outlook and shares in the general prosperity of the valley.

**CARNERO** mining camp, twenty-four miles north of Del Norte, gets its supplies from that town. The principal mines are the Buckhorn, Esperanza the Ada and Spring Chicken. Carnero can be reached via the Denver & Rio Grande railroad and Del Norte. This is a promising camp.

FISHING ON THE RIO GRANDE.

**DEL NORTE** is the oldest town in what is known as the San Juan country. The town-site was surveyed in 1872, though the town company was formed in 1871. At the present time Del Norte has a population of about 1000 souls. The town is reached via the Denver & Rio Grande railroad from Denver or Pueblo—distance from Denver 280 miles. Altitude, 7,400 feet. The town is so situated as to be upon the line between the agricultural and mining sections. To the north and east of the town are the rich and rapidly settling agricultural and pastoral lands of the San Luis Valley; to the south and west are the great mines of San Jaun. Del Norte is beautifully situated in a basin at the foot of the mountain, sheltered from the blasts of winter and having the most delightful weather in summer. The Rio Grande river flows through the edge of the Del Norte town-site, and offers to manufacturing interests the finest water-power in the world. Del Norte has some excellent business and dwelling houses, a fine public school building, two good church buildings—above the average, the Presbyterian College of the Southwest, (a staunch educational institution), a fine flouring mill of the latest roller process, a large brewery using home-grown barley, two banks, a Court House costing $30,000, a weekly newspaper, the United States Land Office, where all business regarding land in this district must be transacted, and countless other enterprises that cannot be mentioned here. A Fair held by the Southwestern Colorado Industrtrial association at Del Norte, in October, 1887, was conceded the most successful effort of the kind ever made in the Southwest. It is the intention of the people to hold a fair, stock show and races every year hereafter. Del Norte has properly been called "The Gateway to the San Juan," being so situated as to control the entrance to the valley of the Rio Grande river, which leads directly to the heart of the San Juan mining country. Del Norte's position is such as to command trade from the mining, agricultural and pastoral sections of the Southwest, and must eventually become a place of great importance. The town of Del Norte is situated on mesa or table land, has all the advantages of good schools and churches, excellent climate, fine water for domestic use—no better anywhere—and is withal an ideal residence and educational center. Del Norte's especial pride is her schools and College. On Lookout Mountain, 600 feet above the town, is mounted a large telescope, to be used in connection with the Presbyterian College of the Southwest. The view from the Lookout observatory is grand in the extreme. The streets of Del Norte are wide, and the town is noted for its growth of trees—mostly cotton-woods. Water for irrigating purposes is supplied by means of a main canal from the Rio Grande river, with laterals over the town-site along the sides of streets. Del Norte has a public reading-room, Grand Army post, lodger of Masons, Knights of Pythias and Good Templars and a company of State militia (the Bowen Rifles).

**GARLAND** is on the New Mexico branch of the Denver & Rio Grande railroad, 325 miles from Denver. It was formerly known as Fort Garland and was a United States military post. Sierra Blanca, elevation 14,464 feet, the highest mountain in the United States with one exception, is 17 miles distant. Good trout fishing and shooting can be found in the adjacent foot hills. Garland's tributary industries are agriculture and stock raising. Enterprising men have invested capital in Garland and are pushing it forward on the road to prosperity.

**LA JARA** is situated in Conejos county, on the New Mexico extension of the Denver & Rio Grande railroad fourteen and a half miles south of Alamosa, and 265 miles from Denver. It is surrounded by a very productive agricultural and pastoral region. Eastern capital has been largely invested in La Jara recently and a rapid improvement is in progress. The future of the town is assured and the growth in population and business has been large and is steadily progressing.

**MONTE VISTA** is located in the midst of 300,000 acres of the richest irrigable land with abundance of water to supply it. It is a new, growing, enterprising, prohibition town, has a superior class of citizens and is beginning to assume city airs. It is rapidly becoming an extra desirable residence locality. It has a first-class roller process flouring mill, 15 stores, 2 banks, planing mill, 3 lumber yards, 3 weekly papers, 3 livery stables, large public library, an $8,000 school house, 7 church organizations, a secular society, secret societies, military company, cornet band, etc. etc. In the vicinity is one farm of 7000 and another of 4000 acres. A $75,000 hotel has been erected, numerous artesian wells are being bored which furnish excellent mineral water, a driving and boating park is established and street car lines chartered. Monte Vista receives and ships great quantities of freight. Its business for 1888 was about $300,-000 and bids fair to double this amount the current year. It has never had a boom, is behind the country, and considering the advantages property is very cheap. Agriculturists with adequate means, manufacturers, business men, miners, health seekers and retired merchants, will find inducements far above the average. There are good openings for many special lines, to wit; Sanitary Hotels, Building Associations, Private Builders with capital, Wholesale Houses, Clothing, Crockery, Boot and Shoe, Hardware and Jewelery stores, Oat Meal Mill, (oats weigh 50 pounds to bushel), Potatoe Starch, Sash and Wagon Factories, Merchant Tailor, Shirt Factory, Sewing Machines, Bakery, Tannery.

Foundry, Creamery, Pork Packing, Brick yards, Flax growing, with rope, paper and oil mills, Nursery, with Wholesale Gardeners, 1000 hop yards, Tobacco Farms, Trotting Horse Breeding, Carp raising, hundreds of health seekers, etc. etc.

**SAGUACHE** is the county seat of Saguache county, and is surrounded by an exceedingly rich agricultural country. Its sources of prosperity are agriculture and stock-raising industries, and it has become firmly established on the best of foundations. It has stages to Villa Grove, Bonanza and Del Norte, and receives its mails daily. It has good business houses, handsome private residence, a large brick school-house, churches and the county buildings.

**SUMMITVILLE** is the name of the most prominent mining camp in the range to the southwest of the valley. It is situated twenty-seven miles from Del Norte. Prominent mills at Summitville are the Annie, Golconda, Aztec & Rio Grande and there are many good mining properties, the Annie, Ida Golconda, Bobtail, Chicago, Esmond, Stars and Stripes and others being among the lot. There is quite a town at Summitville and the mineral output is large and of great value.

**VILLA GROVE** is a prosperous town on the San Luis branch of the Denver & Rio Grande railroad, 247 miles from Denver. It is situated in Saguache county, in the upper or northwestern extremity of the valley. Near it are the mining camps of Kerber Creek, Crestone and Bonanza, where gold and silver are found in paying quantities. Iron is also found near Villa Grove, and the Colorado Coal and Iron Company are working mines of that metal.

**VETERAN** is a town site laid off eighteen miles north of Del Norte, on La Garita. It is favorably located, is in charge of Grand Army men. Around Veteran is some of the best land in the valley. The town can be reached over the Denver & Rio Grande railroad to Del Norte, and from Del Norte by teams over a good road.

# CHAPTER VII

# A Few Condensed Facts.

FOR the convenience of the reader, and to give as much information as possible in a condensed form, the writer has gathered the following paragraphs from authentic sources and grouped them together in this chapter. Each paragraph is complete in itself and contains as accurate a statement of the fact embodied therein, as can be made in the amount of space given.

The mountains surrounding the San Luis Valley, are full of excellent timber, which, for domestic purposes, can be had for the taking. Fine house-logs, fence-poles, bridge-timbers, etc., can be had in this way. Most excellent fire-wood can be had in the same manner.

State lands can be bought or leased of the State by making proper application to the Register of the State Land Board at Denver. These lands, prior to being offered for sale, will be appraised at a minimum price of from $3 to $5 per acre. If these lands are to be leased, they will be appraised at probably $1.25 to $1.50 per acre, and the leasing price per year will be 10 per cent. of the appraised value. Canal lands can be had by application to the agents of the companies in the valley. Improved ranches of 160 acres can be bought in the San Luis Valley at from $1,000 to $5,000 each, according to improvements, location, etc.

Following are approximate prices in the valley on a few articles that may be needed by parties taking up land:

| | |
|---|---:|
| Rough lumber, per 1,000 ft., | $ 20 00 |
| Shingles, per 1,000 | 4 00 |
| No. 8 cook-stove, | 20 00 |
| Lumber wagons, | 95 00 |
| Reversible-share plows, | 12 75 |
| Twelve-inch steel plows, | 13 50 |
| Harrows, | 10 00 |
| Good horses, each, | 100 00 @ 200 00 |
| Cow-ponies, each, | 40 00 @ 100 00 |
| Barbed wire, per 100 pounds, | 6 00 |
| Tarred paper, per cwt., | 4 50 |
| Nails, per keg, | 4 75 |

Farming is carried on in the San Luis Valley mostly by irrigation. This plan insures water every year, drouths or "off years" being unknown in this section. Farming by irrigation, viewed from a scientific stand-point, has many advantages over the old plan of depending upon rains for a water supply. Irrigation is a simple process. The water is conveyed in a ditch along the highest ground in the field to be watered and allowed to flow over the ground as desired. Water purchased of canal companies costs about $1 per acre per year. Permanent water rights can be bought at from $300 to $500 for 160 acres. Stock in irrigating canals, which insures a permanent water-right, can be had at reasonable figures, while there are many entirely private ditches. New canals are being taken out by farmers and others every year. The small grains, vegetables, grasses and hardy fruits grow here to perfection.

Government lands can be had by proper application at the Del Norte Land Office, under the Homestead, Pre-emption, Timber-culture or other laws applying to lands of this class. The Government lands are going rapidly.

Much wool is grown in this country, and there is an excellent opening for a woolen mill, in some one of the thriving towns of the valley.

The roads of this section are generally hard and dry—mud being comparatively unknown here.

In the vally, wells are from ten to thirty feet deep, and generally get good water. In certain localities the water contains some iron, but there is no lime-stone water. "Drive" wells are used to some extent.

While the snow falls very deep on the mountains (usually one to ten feet), there is very little in the valley. It rarely falls over a few inches deep and generally all disappears in a few days.

Alfalfa is raised in large quantities in this section and growns very thrifty. As a forage plant, it has few equals.

During an ordinary winter, much of the stock of the San Luis Valley goes through without a mouthful of feed aside from that picked up on the range. The tendency, however, is toward feeding and a better class of stock.

A single potato weighing two to four pounds is not a rare sight, and 300 to 500 bushels have been raised on an acre. Oats, barley and field-peas do remarkably well, while wheat of excellent quality has been raised in the valley.

The raising of hogs is very profitable in the valley. In the absence of corn they are fattened on peas, which feed is claimed by many ranchmen to be better for fattening purposes than the former grain. White clover is indigenous to the country, and is easily grown.

There are occasional showers of rain during April and May, but the "rainy season" is from about the 6th of July till in September. Then it rains nearly every day, usually in the afternoon, but, when it gets through it quits and gives the sun a chance. There is no such thing as cloudy weather. There are perhaps not over four or five days in the course of a year when the sun does not shine at least part of the day.

Where water is not convenient to the land, and long and expensive canals are required, they are dug by wealthy corporations, who sell water to the ranchmen for irrigating purposes, at about $1 an acre for each season, There are four large ditches of this kind in the San Luis Valley, one of which is sixty feet wide, and extends about twenty-five miles, exclusively of several miles of laterals. Irrigation is a simple process. The water is conveyed in a ditch along the highest ground in the field to be watered, and allowed to flow over the ground as desired.

This valley has been the paradise of the range stock, where they can roam at large and feed upon the best beef-making grasses that nature has prepared, so that they can easily withstand our winters without any feed or attention of any sort. Now the scene will have to change largely, as a smaller number will find pasture in the foot-hills surrounding the valley. The order of the day will now be small herds controlled by inclosures so that the improved breeds can be handled profitably by grazing them in summer, then feed the native and cultivated grasses of the farm in winter, and you can turn out as fine beef as the corn-fed of the States.

Barley produces well in the valley. From forty to sixty bushels per acre is a fair crop. If the proper variety is raised, one can find a ready market for it at $1.15 to $1.25 on cars or at the local brewery. Ground with good oats, it makes strong and healthful feed for horses or other animals.

An average crop of oats is thirty-five to fifty bushels of the finest quality, weighing from forty to forty-eight pounds per bushels. A number of farmers here produced from seventy-five to 100 bushels per acre. The average price is $1.25 to $1.50 per hundred pounds, or 50 to 60 cents per bushel.

# Magnificent Scenery.

**S**) MUCH has been said incidentally in the preceding chapters concerning the scenery of the sunny San Luis that it will not be necessary to take a general view of the valley here. It will, however, serve a good purpose to describe somewhat in detail a few of the more prominent scenic beauties which render this one of the most attractive places in the whole mid-continental region. There are a few people who have not heard of Sierra Blanca, that famous mountain, the highest but one in the United States, whose triple crown dominates the entire Rocky Range and commands the vision from all parts of the valley. It is characterized by the peculiarity of a triple peak and rises directly from the plain to the stupendous height of 14,469 feet, over two miles and three-fifths of sheer ascent. A magnificent view of this mountain is obtained from the cars of the Denver & Rio Grande railroad as soon as the descent from Veta Pass into the San Luis Valley has been made. Surely it is worth a journey across the continent to obtain a view of such a mountain! Although a part of the range, it stands at the head of the valley, like a monarch taking precedence of a lordly retinue. Two-thirds of its height are above timber-line, bare and desolate, and except for a month or two of midsummer, dazzling white with snow, while in its abysmal gorges it holds eternal reservoirs of ice.

> "Oh, sacred mount with kingly crest
> Through tideless ether reaching,
> The earth-world kneels to hear the prayer
> Thy dusky slopes are teaching.
> With mystic glow on sunset eyes
> All trembling lie thy blood-red leaves,
> Their silken veins with gold inwrought.
> Oh, glorious is thy world-wide thought!"

The lower slopes of the mountain are clad in vast forests of pine and hemlock, while its grand triad of gray granite peaks lift into the sky their sharp pyramidal pinnacles, splintered and furrowed by the storm-compelling and omnipotent hand of the Almighty. To the north and south, for a distance of nearly two hundred miles, it is flanked by the serrated crests of the Sangre de Cristo range, the whole forming a panorama of unexampled grandeur and beauty.

A few miles north of Del Norte, and reached by one of the best wagon roads in the country, is the State Park, a pleasure ground set aside by the State for the enjoyment of the people.

WAGON WHEEL GAP.

This is one of the most charming spots in the country and is visited by hundreds of people every season. "The Natural Arch," the "Old Man of Echo Mountain" and scores of other attractions are a source of never-ending interest to the visitors at the place. The State Park can be reached by splendid wagon-roads from Del Norte and is worthy of a visit by sight-seers in this section.

The atmospheric peculiarities of the valley afford some most remarkable spectacles, and a mirage in the summer months is not an infrequent occurrence in the lower portion of the San Luis Valley.

One of these beautiful phenomena is thus described by a spectator of its wonders: First there appeared upon the left and immediate front what looked like a huge sea of heated air, billowy as the ocean. At its northern edge was a long and narrow line of tall and cane-like trees, all at first view covering a strip of country in an east and west line, fully eight or nine miles long. Back of the trees there was for a few moments an unusual disturbance of the heated wave, and then there appeared a beautiful miniature lake, dotted with sailing vessels. But the borders of the lake were extended, and not until the slopes of the northern range formed its borders did it cease to grow. In its growth it extended sufficiently to enclose all the landscape as far as the point of rocks northeast of Hot Springs. Two points attracted more than ordinary attention. That slope of land upon which the Little Corrinne mining claim is located was a sea of molten silver, which reflected from its bosom the mountain with its cragged peaks and stunted pines. An object which rivited the gaze was the form assumed by a tasteful white residence which is so plainly visible. It was not then a house, but a palace moving over a surface of polished silver, while ever and anon appeared trees with wide, spreading branches, which were reflected in the silvery sheen. To the east the beauty was hardly less captivating, as the foothills were transformed into huge Palisades not unlike those to be seen along the Hudson River. The view was most entrancing; its duration was brief, for while the eyes were feasting upon the scene suddenly it vanished. But after all the beauties of the mirage are far surpassed by those of the reality. There are few watering places or health resorts that can vie with Wagon-Wheel Gap, situated thirty miles beyond Del Norte. This place has become the favorite resort for seekers of health and the lovers of the rod and gun. The scenery is wonderfully beautiful. As the Gap is approached the valley narrows until the river is hemmed in between massive walls of solid rock and rise to such a height on either side as to throw the passage into twilight shadow. The river rushes roaring down over gleaming gravel or precipitous ledges. Progressing, the scene becomes wilder and more romantic, until at last the waters of the Rio Grande pour through a cleft in the rocks just wide enough to allow the construction of a road along the river's edge. On the right, as one enters, tower cliffs to a tremendous height, suggestive in their appearance to the Palisades of the Hudson. On the left rises the round shoulder of a massive mountain. The vast wall is unbroken for more than half a mile, its crest presenting an almost unserrated sky-line. Once through the gap, the traveler, look-ng toward the South, sees a valley encroached upon and surrounded by hills

> " Bathed in the tenderest purple of distance,
> Tinted and shadowed by pencils of air."

Here is an old stage station, a primitive and picturesque structure of hewn logs, made cool and inviting by wide-roofed verandas. Not a hun-dred feet away rolls the Rio Grande, swarming with trout. A drive of a mile along a winding road, each turn in which reveals new scenic beauties, brings the tourist to the famous springs. The medicinal quali-ties of the waters, both of the cold and hot springs, have been thoroughly tested and proved equal, if not superior, to the hot springs of Arkansas.

We have briefly sketched the scenic beauties of the valley. Hun-dreds of attractions have been passed by in silence; justice can not be done in the brief space allotted even to the few subjects selected, but if what has been written shall induce the reader to investigate for himself the writer will have the satisfaction of knowing that he will be fully justi-fied in what he has said, and the verdict will be that nothing has been exaggerated concerning the healthfulness, fertility and beauty of the Sunny San Luis.